Bitcoin and Cryptocurrency Simplified for Beginners

Your simple guide to understanding and investing in cryptocurrency

Izumo Tanaka

Table of Content

What Is Bitcoin?

Bitcoin is a digital currency that was created in January 2009. It follows the ideas set out in a white-paper by the mysterious and pseudonymous Satoshi Nakamoto.1 The identity of the person or persons who created the technology is still a mystery. Bitcoin offers the promise of lower transaction fees than traditional online payment mechanisms and, unlike government-issued currencies, it is operated by a decentralized authority.

Bitcoin is a type of cryptocurrency. There are no physical bitcoins, only balances kept on a public ledger that everyone has transparent access to. All bitcoin transactions are verified by a massive amount of computing power. Bitcoins are not issued or backed by any banks or governments, nor are individual bitcoins valuable as a commodity. Despite it not being legal tender, Bitcoin is very popular and has triggered the launch of hundreds of other cryptocurrencies, collectively referred to as alt-coins. Bitcoin is commonly abbreviated as "BTC."

Key Takeaways

Launched in 2009, bitcoin is the world's largest cryptocurrency by market capitalization.

Unlike fiat currency, bitcoin is created, distributed, traded, and stored with the use of a decentralized ledger system, known as a blockchain.

Bitcoin's history as a store of value has been turbulent; the cryptocurrency skyrocketed up to roughly $20,000 per coin in 2017, but less than years later, it was trading for less than half of that.

As the earliest virtual currency to meet widespread popularity and success, bitcoin has inspired a host of other cryptocurrencies in its wake.

Understanding Bitcoin

The bitcoin system is a collection of computers (also referred to as "nodes" or "miners") that all run bitcoin's code and store its blockchain. Metaphorically, a blockchain can be thought of as a collection of blocks. In each block is a collection of transactions. Because all the computers running the blockchain has the same list of blocks and transactions, and can transparently see these new blocks being filled with new bitcoin transactions, no one can cheat the system.

Anyone, whether they run a bitcoin "node" or not, can see these transactions occurring live. In order to achieve a nefarious act, a bad actor would need to operate 51% of the computing power that makes up bitcoin. Bitcoin has around 12,000 nodes, as of January 2021, and this number is growing, making such an attack quite unlikely.

But in the event that an attack was to happen, the bitcoin miners—the people who take part in the bitcoin network with their computer—would likely fork to a new blockchain making the effort the bad actor put forth to achieve the attack a waste.

Balances of bitcoin tokens are kept using public and private "keys," which are long strings of numbers and letters linked through the mathematical encryption algorithm that was used to create them. The public key (comparable to a bank account number) serves as the address which is published to the world and to which others may send bitcoins.

The private key (comparable to an ATM PIN) is meant to be a guarded secret and only used to authorize bitcoin transmissions. Bitcoin keys should not be confused with a bitcoin wallet, which is a physical or digital device that facilitates the trading of bitcoin and allows users to track ownership of coins. The term "wallet" is a bit misleading, as bitcoin's decentralized nature means that it is never stored "in" a wallet, but rather decentrally on a blockchain.

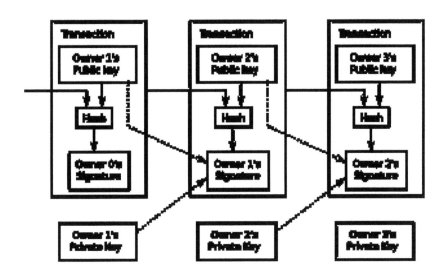

Peer-to-Peer Technology

Bitcoin is one of the first digital currencies to use peer-to-peer technology to facilitate instant payments. The independent individuals and companies who own the governing computing power and participate in the bitcoin network—bitcoin "miners"—are in charge of processing the transactions on the blockchain and are motivated by rewards (the release of new bitcoin) and transaction fees paid in bitcoin.

These miners can be thought of as the decentralized authority enforcing the credibility of the bitcoin network. New bitcoin is released to the miners at a fixed, but periodically declining rate. There are only 21 million bitcoin that can be mined in total. As of January 30, 2021, there are approximately 18,614,806 bitcoin in existence and 2,385,193 bitcoin left to be mined.

In this way, bitcoin other cryptocurrencies operate differently from fiat currency; in centralized banking systems, currency is released at a rate matching the growth in goods; this system is intended to maintain price stability. A decentralized system, like bitcoin, sets the release rate ahead of time and according to an algorithm.

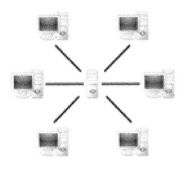

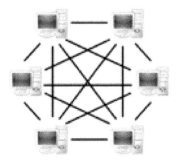

Server-based   P2P-network

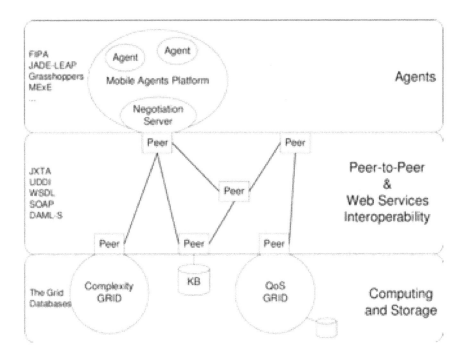

Bitcoin Mining

Bitcoin mining is the process by which bitcoins are released into circulation. Generally, mining requires the solving of computationally difficult puzzles in order to discover a new block, which is added to the blockchain.

Bitcoin mining adds and verifies transaction records across the network. For adding blocks to the blockchain, miners are rewarded with a few bitcoins; the reward is halved every 210,000 blocks. The block reward was 50 new bitcoins in 2009. On May 11th, 2020, the third halving occurred, bringing the reward for each block discovery down to 6.25 bitcoins.

A variety of hardware can be used to mine bitcoin. However, some yield higher rewards than others. Certain computer chips, called Application-Specific Integrated Circuits (ASIC), and more advanced processing units, like Graphic Processing Units (GPUs), can achieve more rewards. These elaborate mining processors are known as "mining rigs."

One bitcoin is divisible to eight decimal places (100 millionths of one bitcoin), and this smallest unit is referred to as a Satoshi. If necessary, and if the participating miners accept the change, bitcoin could eventually be made divisible to even more decimal places.

History of Bitcoin

After the subprime crisis and the Lehman Brothers bankruptcy shedding light on vulnerabilities of the financial system in place has been published the 31st October 2018 the Bitcoin white paper detailing how the protocol works. This new exchange system has learned from the failed but instructive experiences of its predecessors HashCash (1997), B-Money (1998), RPOW (2004) and Bitgold (2008).

Bitcoin echoed, in particular, the Cypherpunks, a community that defended privacy rights and had detected well in advance the problems that the Internet would pose in terms of privacy breaches. Satoshi Nakamoto, a pseudonym used by the creator(s), was able to surround himself with members of this community who had already worked on the projects mentioned above such as Nick Szabo, Hal Finney, Adam Back, and Wei Dai.

Let us take again the important points of the history of Bitcoin in a chronological way since its installation :

2009

January 3: Satoshi Nakamoto launches the platform by generating the first Bitcoin block, also called "genesis block". 9 days later, the first transaction took place in block 170 between Satoshi Nakamoto and Hal Finney for 10 BTC. For the anecdote, the block generated includes the text: "The Times 03/Jan/2009 Chancellor on brink of second bailout for banks", Satoshi proves that he did not mine a block before January 3, 2009.

October 5th: the first exchange rate is given on the New Liberty Standard platform. It is 1309.03 BTC for 1$, this price is established by calculating the price of the electricity spent to undermine this quantity of bitcoins.

2010

May 22: Programmer Lazslo Hanyecz makes the first Bitcoin transaction against a property. It was two pizzas for a price of 10000 BTC, equivalent to 20$ at market price.

July 17: MtGox is born. This collecting card exchange will become the largest Bitcoin trading place for the next 4 years.

December 12: Satoshi Nakamoto disappears from the project and gives the lead to Gavin Anderson, a developer still active today on Bitcoin.

2011

February: Bitcoin reaches parity with the dollar.

The famous illicit marketplace Silk Road is open and uses Bitcoin as a means of payment.

The first crypto asset emerging as alternatives to Bitcoin. Namecoin and Litecoin were the first to present themselves.

2012

27 September: Creation of the Bitcoin Foundation, which aims to standardize, protect and promote Bitcoin.

28 November: the first "Halving" takes place, dividing the reward awarded to minors with each block of transactions by two. It is now 25 BTC.

2013

October 2: Silk Road closes. The price of Bitcoin will begin a new period of progress: it surpasses its previous record of 237 euros on 7 November to achieve 912 euros on December 4.

2014

February: The world's largest marketplace MtGox goes bankrupt following a hack of 744,000 BTC, resulting in a colossal drop in the price of Bitcoin that will last until 2015.

During this period, the course collapses more than 750 euros to go below 160 euros.

May 13: the birth of La Maison du Bitcoin, the first and only space dedicated to cryptoassets in France, bringing expertise and know-how to the general public, and also allowing easy and serene buying and selling via its Coinhouse platform.

2015

Groups such as UBS, IBM, Orange and the American army are beginning to take Bitcoin and blockchain technologies seriously and are working on use cases.

Development of the Ethereum project led by Vitalik Buterin, which will develop to become the second crypto-active in terms of market capitalization. Ethereum offers a decentralized platform facilitating the creation of decentralized applications, called Apps.

2016

August 2: Bitfinex suffers one of the biggest hacks in the history of Bitcoin with a 119 756 BTC theft.

The crypto ecosystem is growing significantly with a strong increase in the number of projects and a stable and gradual increase in the price of cryptoassets.

2017

On 4 April, Japan introduced a legislative framework for "virtual currencies" and recognized them as legal means.

The explosion of the courses and the craze around the crypto assets made rise the price of a Bitcoin until 20000$. This digital gold rush has also shown the current limitations of technology as a global means of payment, with transaction fees of up to $30 per transaction.

However, the community has proven its ingenuity and efficiency by proposing solutions like SegWit and the Lightning Network that will drastically increase the capacity of the Bitcoin network.

2018

This year will remain in investors' minds the year of the "bear market" with a strong correction of Bitcoin to the extend of 85% of its ATH.

2018 is also the year of the institutionalization of the market, with regulated investment funds such as Greyscale welcoming more and more professional investors wishing to expose themselves to Bitcoin.

A collective awareness is also emerging after 2017, with regulators and governments educating themselves on Bitcoin and Blockchain networks in general, it is a year for learning.

The Lightning Network was launched in early 2018, and is growing continuously throughout the year.

2019

2019 will mark the end of the bear market, with a strong recovery in the price of bitcoin from the beginning of the year.

Bitcoin reaches new heights in terms of computing power dedicated to its security as well as in terms of use.

Who controls Bitcoin?

Bitcoin is the creation of a person or group of people hiding behind the pseudonym "Satoshi Nakamoto". He (or she) has been able to merge different technologies combining cryptography and distributed registers in order to offer a valuable network without trusted third parties. Despite the success of his invention, Satoshi's identity remains unknown, and he quit the project in 2011. But if Satoshi came back tomorrow, would he be able to control Bitcoin? Is the protocol really secure?

Speculation about Satoshi Nakamoto's true identity has been recurrent since Bitcoin's early years. Satoshi Nakamoto's reasons for remaining anonymous are unknown, although its is likely he wished to avoid significant media harassment and potential legal proceedings, as Bitcoin has been banned in several countries. Satoshi Nakamoto left the project in 2011, leaving Bitcoin in the hands of its users and thus fulfilling the vision of a decentralised and censorship resistant network of value: the emergence of a truly different currency.

How can we be sure that Satoshi Nakamoto is not benefiting from his invention?

Bitcoin is an open source software with all the code available on the dedicated github page. The study of the software's code allows us to know precisely how it behaves. This makes it easy to verify that Satoshi could under no circumstances manipulate the Bitcoin protocol and make it evolve at its own discretion or change the content of a particular portfolio. Based on promising and disruptive technology, we believe that Bitcoin has many advantages and that's why you can buy bitcoins on our online platform.

Who controls the protocol?

There are many answers to this question. They are very well detailed here by engineer Jameson Lopp.

In Bitcoin, everything is designed and configured to avoid centralisation and single points of failure. You can read our article on the 21 million bitcoins limit, which is an important feature of the protocol.

While everyone can contribute to the code on Github/bitcoin, a number of safeguards are implemented to ensure that the code and its evolutions remain secure.

A limited number of individuals, the core developers, have the ability to publish and sign a new version of the protocol using what is called their PGP key: the equivalent of a Bitcoin private portfolio key. Today, five people hold verified keys. This number is dynamic and these people are known to have been actively involved in the development of the protocol for many years. They may choose to add/remove a member at any time by consensus.

Any other developer has the ability to propose changes to the protocol, and core developers are responsible for reading them and accepting or rejecting them.

So the protocol is in the developers hands?

First of all, there are several implementations of the Bitcoin software. While Bitcoin Core is the main implementation on which most developers work with more than 95% of nodes, there is no requirement to use it. The application being compatible with the protocol is enough, and it will be usable throughout the network, in the same way that web users have the choice between Chrome, Firefox or Edge.

But even if you use a different software, it must comply with the protocol. However, the protocol is also controlled by core developers. Can we then still talk about decentralisation, since it would be enough for some of them to agree to modify it in the way they would like?

The principle of consensus

Fortunately not: the only power core developers hav is to publish and propose new versions of the protocol to the community. Ultimately, the owners of the network nodes are responsible for deciding whether or not to update their machines with a new version. It can therefore be said that the decision to change the protocol is taken by consensus by the owners of the nodes of the Blockchain.

Nothing prevents a new development team from emerging and proposing a new version. If it is accepted by the vast majority of nodes, it becomes the official version.

This freedom was used in 2017 for the implementation of SegWit. The miners were reluctant to adopt SegWit. A proposal was then made under the name BIP148 to also involve node owners and give their opinions. As this proposal was not applied to Bitcoin Core, an implementation called Bitcoin UASF quickly gained popularity. The pressure put by Bitcoin UASF was enough for the miners to integrate SegWit via the processes defined in Bitcoin Core.

In a nutshell, the technology and protocol that make Bitcoin work allows it to evolve dynamically, on the principle of consensus and not unilateral decision. The answer to the question "Who controls Bitcoin?" is simple: "nobody and everyone at the same time", through a mechanism not based on democracy, but on consensus. Welcome to the world of free software.

Can Bitcoin be hacked?

A strong paradigm in Bitcoin is based on its supposedly absolute security, with its immutable and unhackable blockchain. Yet, we often hear about hackers in the Bitcoin world, with sometimes hundreds of millions of dollars stolen. What is it really like?

Hacking? Where?

During 2018, there were numerous attacks on Bitcoin or other cryptoassets. The year started off very badly, with some NEM tokens stolen on a Japanese platform for more than $500 million. Unfortunately the lessons were not learned, as more than $1 billion was stolen during the year on multiple platforms...

But the most widely reported hack of cryptoassets is undoubtedly the one suffered by the MtGox platform at the beginning of 2014, which made headlines by reaching the news of most countries in Europe and the United States, and which involved an amount of 700,000 bitcoins.

These attacks have one thing in common: they exclusively concern online platforms on which investors store their cryptoassets.

What is the difference between a portfolio and a platform?

Storing Bitcoins, as we have already mentioned, is a misleading term. All the bitcoins, as well as the other cryptoassets, only exist on the Blockchain. The owner of a bitcoin actually only keeps a private key, the purpose of which is to authorise transactions of this specific bitcoin.

In the case where a person uses portfolios such as Coinomi or the Ledger Wallet, they directly control their private keys to spend the bitcoins. A potential hacker will therefore be forced to obtain the private keys of the portfolio, which may be difficult or even impossible, depending on the portfolio chosen, for a result that is at best random.

An online platform is of course a much more profitable target. These platforms handle millions of dollars worth of bitcoins and other cryptoassets on a daily basis, and they operate via standard computer servers. Investors put their Bitcoins on these platforms, which therefore possess the private keys to spend them. A hacker who gets his hands on these private keys is immediately able to recover all the bitcoins stored on the platform.

Another form of attack consists in hacking the platform's removal system to make it believe in a legitimate cryptographic withdrawal, thus recovering funds belonging to many users.

Platforms are hacked, not the Blockchain

Hackers focus on the security of online platforms, whose software has been written by the employees of these platforms. The software that defines the Bitcoin protocol and the Blockchain, on the other hand, was only found to be defective once, in 2010, and the fault was quickly fixed, with no financial consequences.

One of the fundamental reasons that can explain the solidity of the protocol is the fact that it is open source: it is readable by anybody who want it. It may seem as a paradox to open such a software, but experience has shown that free softwares such as the Linux operating system is extremely stable and secure, with a large community working together on a daily basis to improve its quality and resolve any security problems.

Bitcoin is no exception to this rule, and a community of more than a hundred IT specialists is working to secure and develop it everyday.

So it is wrong to say that the many hacks that take place on online trading platforms question the security of Bitcoin or the Blockchain. It is only the software on these platforms that is defective and not the protocol itself. To make an analogy, it can be said that just because a bank is robbed does not mean that the security mechanisms for euro banknotes are defective.

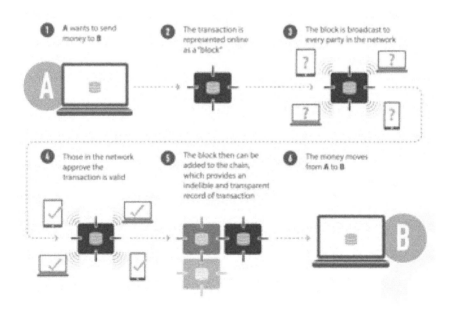

What determines the price of Bitcoin?

If a Bitcoin were a manufactured product owned by a company, it would be possible for its creator to set the price, either directly by arbitrarily determining it or indirectly by limiting or increasing the quantity available on the market. As Bitcoin is by nature decentralised, it is not possible to use these two methods to control its price.

As we have seen previously, the number of new bitcoins generated is fixed over a certain period of time by the mining process. New bitcoins are "distributed" depending on the computing power on the network (hashrate).

Bitcoin, a volatile asset

Every day, the price of Bitcoin and other cryptoassets tends to fluctuate significantly. It is not uncommon to see increases or decreases of 10% or more over a day, which is extremely rare for other asset classes such as equities or precious metals.

Like these other assets, the value of Bitcoin is determined by supply and demand in marketplaces. At any given time, economic actors decide to buy and sell at a price they consider interesting. When a buyer and a seller agree on a price, a transaction is made and the price is set according to this last transaction.

Fundamental analysis

The analysis of a project, its technological and economic components, its management by the development team, are excellent factors in determining the value of an asset such as Bitcoin. We can add new ones, for example a large company using the project technology, which can develop its adoption. "The analysis of the technological and economic components of a project is an excellent factor to determine the value of an asset such as Bitcoin"

However, the valuation of the crypto market is difficult to determine because the underlying is not well defined. The value of a share is strongly influenced by the results of the company to which it corresponds. This is one of the reasons why Coinhouse offers you a personalised program to invest in this market.

Technical analysis

Technical analysis consists in analyzing the chart curves of the various assets in order to identify trends and predict future developments based on past market movements.

Some cartesian minds will consider it to be saying that it will rain next Wednesday because it rained on the previous three Wednesdays. But when the majority of economic actors agree on the validity of such indicators, the market can indeed react in the expected direction.

"Technical analysis consists in analyzing the chart curves of the various assets in order to identify trends and predict future market developments"

What Is Crypto Mining? How Cryptocurrency Mining Works

Although crypto mining has only been around since Bitcoin was first mined in 2009, it's made quite a splash with miners, investors and cybercriminals alike. Here's what to know about cryptocurrency mining and how it works…

Crypto mining (or "cryptomining," if you'd prefer) is a popular topic in online forums. You've probably seen videos and read articles about Bitcoin, Dash, Ethereum, and other types of cryptocurrencies. And in those pieces of content, the topic of cryptocurrency mining often comes up. But all of this may leave you wondering, "what is Bitcoin mining?" or "what is crypto mining?"

In a nutshell, cryptocurrency mining is a term that refers to the process of gathering cryptocurrency as a reward for work that you complete. (This is known as Bitcoin mining when talking about mining Bitcoins specifically.) But why do people crypto mine? For some, they're looking for another source of income. For others, it's about gaining greater financial freedom without governments or banks butting in. But whatever the reason, cryptocurrencies are a growing area of interest for technophiles, investors, and cybercriminals alike.

What Is Crypto Mining? Cryptocurrency Mining Explained

The term crypto mining means gaining cryptocurrencies by solving cryptographic equations through the use of computers. This process involves validating data blocks and adding transaction records to a public record (ledger) known as a blockchain.

In a more technical sense, cryptocurrency mining is a transactional process that involves the use of computers and cryptographic processes to solve complex functions and record data to a blockchain. In fact, there are entire networks of devices that are involved in crypto-mining and that keep shared records via those blockchains.

It's important to understand that the cryptocurrency market itself is an alternative to the traditional banking system that we use

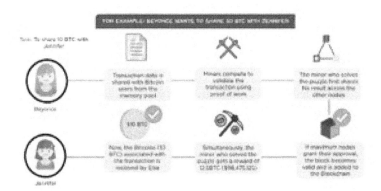

globally. So, to better understand how crypto mining works, you first need to understand the difference between centralized and decentralized systems.

Traditional Banks Are Centralized Systems

In traditional banking, there's a central authority that controls, maintains, and updates a centralized record (ledger). That means that every single transaction has to go through the central banking system, where it's recorded and verified. Plus, it's a restricted system — only a small number of organizations (banks) are allowed to connect to the centralized banking system directly.

Cryptocurrencies Use Decentralized, Distributed Systems

With cryptocurrencies, there's no central authority, nor is there a centralized ledger. That's because cryptocurrencies operate in a decentralized system with a distributed ledger (more on this shortly) known as blockchain. Unlike the traditional banking system, anybody can be directly connected to and participate in the cryptocurrency "system." You can send and receive payments without going through a central bank. That's why it's called decentralized digital currency.

But in addition to being decentralized, cryptocurrency is also a distributed system. This means the record (ledger) of all transactions is publicly available and stored on lots of different computers. This differs from the traditional banks we mentioned earlier, which are centralized systems.

But without a central bank, how are transactions verified before being added to the ledger? Instead of using a central banking system to verify transactions (for example, making sure the sender has enough money to make the payment), cryptocurrency uses cryptographic algorithms to verify transactions.

And that's where bitcoin miners come in. Performing the cryptographic calculations for each transaction adds up to a lot of computing work. Miners use their computers to perform the cryptographic work required to add new transactions to the ledger. As a thanks, they get a small amount of cryptocurrency themselves.

Understanding the Terms: Centralized, Decentralized, and Distributed

To help you better understand what I'm talking about, let's consider the following graphic:

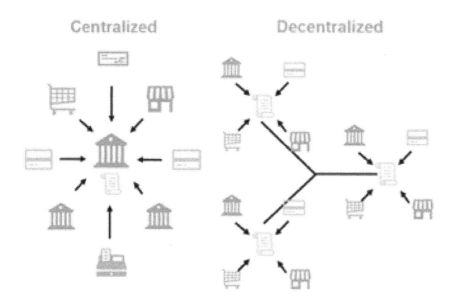

The examples in the graphic above display the differences between a centralized system and a decentralized one.

In the left half of the graphic is an illustration of a centralized system. The traditional centralized currency system in the U.S. operates through the use of computers, networks and technologies that are owned, operated and maintained by financial institutions. So, whenever you send money to a family member or a friend, that transaction goes through your bank.

half of the graphic), operates using a network of separately owned, operated and maintained devices. They lend their resources to create this decentralized network and share the responsibility of verifying transactions, updating and maintaining redundant versions of the ledger simultaneously.

Is Crypto Mining Legal?

In general, the answer is yes. Determining whether crypto mining is legal or illegal primarily depends on two key considerations:

Your geographic location, and

Whether you mine crypto through legal means.

However, where you start to tread into the territory of illegal activities is when you use illicit means to mine cryptocurrencies. For example, some cybercriminals use Javascript in browsers or install malware on unsuspecting users' devices to "hijack" their devices' processing power. This type of cyber attack is known as cryptojacking. We're going to publish a separate article on that topic later this month, so stay tuned.

But it's important to note that cryptocurrency mining is viewed differently by various governments around the globe. The U.S.

Library of Congress published a report stating that in Germany, for example, mining Bitcoin is viewed as fulfilling a service that's at the heart of the Bitcoin cryptocurrency system. The LOC also reports that many local governments in China are cracking down on Bitcoin mining, leading many organizations to stop mining Bitcoin altogether.

Furthermore, some countries view cryptocurrency mining profits as being taxable while other countries view the fruits of such activities as non-taxable income.

We'll talk more about what makes cryptocurrencies and crypto mining so appealing in a bit. But first, let's break down how cryptocurrency mining actually works. To do this, we'll explore the technologies and processes that are involved in it.

How Cryptomining Works (And an In-Depth Look at Blockchain)

In a nutshell, crypto miners verify the legitimacy of transactions in order to reap the rewards of their work in the form of cryptocurrencies. To understand how most cryptocurrency mining works in a more technical sense, you first need to understand the technologies and processes behind it. This includes understanding what blockchain is and how it works.

The first thing to know is that two things are central to the concept of blockchain: public key encryption and math. While I'm definitely a fan of the first, I'll admit that the latter isn't my strong suit. However, public key cryptography (aka public key encryption or asymmetric encryption) and math go together in blockchains like burgers and beer.

Traditional cryptocurrencies such as Bitcoin use a decentralized ledger known as blockchain. A blockchain is a series of chained data blocks that contain key pieces of data, including cryptographic hashes. These blocks, which are integral to a blockchain, are groups of data transactions that get added to the end of the ledger. Not only does this add a layer of transparency, but it also serves as an ego inflator when people get to see their transactions being added (chained) to the blockchain. Even though it doesn't have their names listed on it, it often still evokes a sense of pride and excitement.

Breaking Down the Roles and Processes Within the Bitcoin Blockchain

There are several key components and processes involved in the creation of a blockchain. For this explanation, we're going to use

Bitcoin as our example:

Nodes. These are the individuals and devices that exist within the blockchain (such as your computer and the computers of other cryptocurrency miners).

Miners are the specific nodes whose jobs are to verify ("solve") unconfirmed blocks in the blockchain by verifying the hashes. Once a miner verifies a block, the confirmed block then gets added to the blockchain. The first miner who announces to the rest of the nodes that they've solved the hash is rewarded with a cryptocurrency.

Transactions. A transaction is the thing that gets this party started — I mean, the cryptocurrency mining process rolling. To put it simply, a transaction is an exchange of cryptocurrencies between two parties. Each separate transaction gets bundled with others to form a list that gets added to an unconfirmed block. Each data block must then be verified by the miner nodes.

Hashes. These one-way cryptographic functions are what make it possible for nodes to verify the legitimacy of cryptocurrency mining transactions. A hash is an integral component of every block in the blockchain. A hash is generated by combining the header data from the previous blockchain block with a nonce.

Nonces. A nonce is crypto-speak to describe a number that's used only once. Basically, NIST describes a nonce as "a random or non-repeating value." In crypto mining, the nonce gets added to the hash in each block of the blockchain and is the number that the miners are solving for.

Consensus algorithm. This is a protocol within blockchain that helps different notes within a distributed network come to an agreement to verify data. The first type of consensus algorithm is thought to be "proof of work," or PoW.

Blocks. These are the individual sections that compromise each overall blockchain. Each block contains a list of completed transactions. Blocks, once confirmed, can't be modified. Making changes to old blocks means that the modified block's hash — and those of every block that's been added to the blockchain since that original block was published — would then have to be recognized by all of the other nodes in the peer-to-peer network. Simply put, it's virtually impossible to modify old blocks.

Blockchain. The blockchain itself is a series of blocks that are listed in chronological order. Because previously published blocks can't be modified or altered after they've been added to the blockchain, this provides a level of transparency. After all, everyone can see the transactions.

A Step-by-Step Look at the Crypto Mining Process

Okay, it's time to take a really granular look at the cryptocurrency mining process and better understand how it works.

1. Nodes Verify Transactions Are Legitimate

Transactions are the basis that a cryptocurrency blockchain is built upon. So, let's consider the following example to understand how this all comes together:

Let's say you're a crypto miner and your friend Andy borrows $5,000 from your other friend Jake to buy a swanky new high-end gaming setup. It's a top-of-the-line computer that's decked out with the latest gaming setup accoutrements. (You know, everything from the LED keyboard and gaming mouse to the wide multi-screen display and killer combo headset with mic.) To pay him back, Andy sends him a partial Bitcoin unit. However, for the transaction to complete, it needs to undergo a verification process (more on that shortly).

2. Separate Transactions Are Added to a List of Other Transactions to Form a Block

The next step in the crypto mining process is to bundle all

transactions into a list that's then added to a new, unconfirmed block of data. Continuing with the example of the gaming system transaction, Andy's Bitcoin payment to Jake would be considered one such transaction.

By adding their transaction to the blockchain (once the verification process is complete), it prevents "double spending" of any cryptocurrencies by keeping a permanent, public record. The record is immutable, meaning it can never be manipulated or altered.

3. A Hash and Other Types of Data Are Added to the Unconfirmed Block

Once enough transactions are added to the block, additional info is added as well, including the header data and hash from the previous block in the chain and a new hash for the new block. What happens here is that the header of the most recent block and a nonce are combined to generate the new hash. This hash gets added to the unconfirmed block and will then need to be verified by a miner node.

In this case, let's say you're just lucky enough to be the one to solve it. You send a shout-out to all of the other miners on the network to say that you've done it and to have them verify as much.

4. Miners Verify the Block's Hash to Ensure the Block Is Legitimate.

In this step of the process, other miners in the network check the veracity of the unconfirmed block by checking the hash.

But just how complex is a hash? As an example, let's imagine you apply a SHA-256 hash to the plain text phrase "I love cryptocurrency mining" using a SHA-256 hash calculator. This means that the phrase would becomes "6a0aa6e5058089f590f9562b3a299326ea54dfad1add8f0a141b731 580f558a7." Now, I don't know about you, but I'm certainly not going to be able to read or decipher what the heck that long line of ciphertext gibberish says.

5. Once the Block is Confirmed and the Block Gets Published in the Blockchain

On the crypto miner's side of things, this is the time for celebration because the proof of work (PoW) is now complete. The PoW is the time-consuming process of solving the hash and proving to others that you've legitimately done so in a way that they can verify.

From the user's side of things, it basically means that Andy's transfer of a partial Bitcoin to Jake is now confirmed and will be added to the blockchain as part of the block. Of course, as the most recently confirmed block, the new block gets inserted at the end of the blockchain. This is because blockchain ledgers are chronological in nature and build upon previously published entries.

How These Components Work Together in the Blockchain Ecosystem

So, how does this ledger stay secure from manipulation and unauthorized modifications? All of the transactions for the ledger are encrypted using public key cryptography. For the blocks to be accepted, they must utilize a hash that the miner nodes on the blockchain can use to verify each block is genuine and unaltered.

Who Updates the Blockchain (and How Frequently)?

Because there's no centralized regulating authority to manage or control exchanges, it means that the computers that mine that specific type of cryptocurrency are all responsible for keeping the ledger current. And updates to the blockchain are frequent. For example, Buybitcoinworldwide.com estimates that the Bitcoin blockchain gains a new block every 10 minutes through the mining process.

With a cryptocurrency blockchain, anyone can see and update the ledger because it's public. You do this by using your computer to generate random guesses to try to solve an equation that the blockchain system presents. If successful, your transaction gets added to the next data block for approval. If not, you go fish and keep trying until either you're eventually successful. Or you decide to spend your time and resources elsewhere.

Now that you understand what cryptocurrency mining is and how it works, let's take a few moments to understand the attraction of cryptocurrencies and why someone would want to mine them.

When is the best time to buy cryptocurrencies?

Finding the perfect entry price for an investment to guarantee the best return has always been the equivalent of the speculators' quest for the Holy Grail. Many profitable investment opportunities were missed in order to buy "cheaper". Many potential profits often disappeared for speculators who wanted to sell "at higher prices". Should solutions such as dollar cost averaging be considered?

Trading, a different job than investing

If day-trading may seem attractive, promising the possibility of earning fortunes by staying in front of your computer all day, the reality is quite different. According to Forbes, 90% of day-traders end their year in the red.

Most of the remaining 10% have made trading their profession with a constant watch on markets and stock fundamentals. It is a complex, stressful and risky job that requires real skills and the ability to remain calm at all times.

Contrary to what "training" sellers, youtubers and other

"influencers" often want to make you believe, trading can very hardly be an additional activity to another occupation.

There are only two ways to make money: the first is through work, the second is by taking risks. Trading is both a lot of work and a lot of risk

The volatility of Bitcoin and its role as a medium of exchange and a store of value

Bitcoin is among the most volatile assets, with the cryptocurrency often experiencing extreme price swings in short periods.

While a high level of volatility opens up opportunities for traders, purchasing and holding the asset also comes with increased risks.

A highly volatile asset presents increased risks to investors as it is more prone to major price swings in short periods.

Investors could take advantage of an asset that has high levels of volatility to make more (potential) profits.

Along with the energy industry, commodities, and emerging currencies, cryptocurrency is among the most volatile asset classes.

Bitcoin's volatility is influenced by several factors, including the size of the market, the liquidity of the asset, the impact of news, regulation, and the extent to which crypto investors act on speculation.

With a pro-regulation approach from governments and innovation-focused mindset from businesses within the sector, the cryptocurrency market could continue to grow organically, potentially reaching mainstream adoption in the coming years.

What Is Volatility?

Let's start with the basics before we deep-dive into exploring Bitcoin's volatility.

Volatility is a statistical measure indicating the dispersion of returns for a specific asset (e.g., Bitcoin, stocks, bonds), or market index (e.g., the S&P 500, NASDAQ 100).

In layman terms, higher volatility means larger price swings (in any direction), while lower levels of volatility represent more stable and predictable price levels for an asset.

Volatility is one of the most important factors to determine the risk level of different assets for investors. The higher the volatility, the more risks an asset poses for traders.

It's always what makes Bitcoin so attractive.

In one of our previous articles, we studied in depth how Bitcoin acts in comparison to gold, where we found that while gold has an annual volatility rate of 10%, Bitcoin stands out with a staggering 95%.

What Are the Most Volatile Asset Classes?

Some assets are more volatile than others.

As you can see in the chart above, Bitcoin is among the most volatile assets out there.

What's even more interesting is that – despite being highly volatile – BTC showcases the lowest levels of volatility among digital assets.

And it makes sense as Bitcoin has the largest market capitalization, an already established infrastructure, a strong community, as well as a solid reputation as the world's original cryptocurrency.

The energy industry – which includes assets like oil, gas, coal, and renewable energy technologies – features the highest volatility in global finance. Over the past few years, oil's volatility has even exceeded Bitcoin's multiple times.

As a result of the current COVID-19 outbreak and the Russia–Saudi Arabia oil price war, the asset's volatility surged in a very short period, being 2.5 times more volatile than Bitcoin at the time of writing this article.

The commodity sector – which includes natural resources like oil and gas, precious metals, and agricultural goods like beef and grain – often feature high levels of volatility.

Emerging currencies represent the national currencies of countries that are in the process of economic development.The leaders in this category are often referred to as BRIC (Brazil, Russia, India, and China).

Emerging currencies are much more volatile than major fiat currencies (e.g., USD, EUR, GBP).

In fact, the least volatile asset classes include major fiat currencies like the USD and EUR, low-yield treasury bonds of developed countries (e.g., the United Kingdom, Germany), low-volatility exchange-traded funds (ETFs), and stocks of established companies in low-risk sectors.

How to Calculate Bitcoin's Volatility?

While there are multiple ways to measure volatility, the most widely accepted method indicates the standard deviation between returns from the same market asset or index.

You can either calculate Bitcoin's volatility yourself or use pre-calculated BTC volatility indexes using external resources, such as BitPremier or Woobull.

Example Calculation

Let's say that we would like to measure the volatility of Bitcoin for a 12 month period.

For simplicity, let's say that BTC closed the first month with a price of $1,000, the second month with $2,000, and then continuing this $1,000 monthly value increase until the end of the period.

Take the following steps to calculate Bitcoin's volatility for this period using the standard deviation model:

The first step is to find the mean or average price of Bitcoin for the period. The easiest way to do this is to add together the values of each month and divide them by the total number of months:

Example: $1,000 + $2,000 + $3,000 +... + $12,000 = $78,000 / 12 months = $6,500

The next step is to calculate the difference between the BTC closing price for each month and the mean price for the period. As we need each value, we recommend using a spreadsheet to calculate this stat (also called deviation).

Example: $12,000 – $6,500 = $5,500

To eliminate negative values, each deviation value should be squared.

Example: $(-5,500)2 = 30,250,000$

When you have all the deviation values for each month, add them together.

Example: 30,250,000 + 20,250,000 + ... + 30,250,000 = 143,000,000

Now divide the sum of the squared deviation values by the total number of months to calculate the variance.

Example: 143,000,000 / 12 = 11,916,667

Take the square root of the variance you just calculated to measure the standard deviation of Bitcoin's price for 12 months.

Example: $\sqrt{11,916,667} = 3,452$

	BTC Price	Mean Price	Deviation	Squared Deviation	Variance	Standard Deviation
Month 1	1,000	6,500	-5,500	30,250,000		
Month 2	2,000	6,500	-4,500	20,250,000		
Month 3	3,000	6,500	-3,500	12,250,000		
Month 4	4,000	6,500	-2,500	6,250,000		
Month 5	5,000	6,500	-1,500	2,250,000		
Month 6	6,000	6,500	-500	250,000		
Month 7	7,000	6,500	500	250,000		
Month 8	8,000	6,500	1,500	2,250,000		
Month 9	9,000	6,500	2,500	6,250,000		
Month 10	10,000	6,500	3,500	12,250,000		
Month 11	11,000	6,500	4,500	20,250,000		
Month 12	12,000	6,500	5,500	30,250,000		
Total	78,000			143,000,000	11,916,667	3,452

The standard deviation of $3,452 shows how values are spread out around the average Bitcoin price. Use this value to gather insight into how far the BTC price may deviate from the average value of the digital asset.

What Factors Influence Bitcoin's Volatility?

As mentioned before, cryptocurrency is a highly volatile asset class that is prone to major price swings in short periods, but why?

An Emerging Asset Class

With 2009 marking the birth of Bitcoin, the crypto market is still very young.

And history has proven that an infant market that features new technology is subject to increased levels of volatility and turbulence.

New technology always presents an increased risk for investors as the failure rate is higher at the start.

At the time of writing this article, the total market capitalization of the crypto market is approximately $204 billion.

In contrast, Microsoft (MSFT) stocks alone have a market cap of $1.32 trillion, which is nearly 6.5 times larger than the whole digital asset sector's.

The difference is even higher between the market capitalization of the S&P 500 ($21.42 trillion) and the crypto industry, with the prior representing a 105 times bigger market cap than digital assets.

These stats obviously show that digital assets have not yet been adopted by the masses.

However, as time passes, we can expect this technology to further develop, and investors in crypto would argue that as more people come to adopt crypto for usage (instead of mere speculation), we can expect market cap to go up and volatility to go down.

Example: Tech Stocks and the Dot Com Bubble

Let's look at a past example of a relatively new sector that has been highly volatile.

The late 1990s and early 2000s are known for the rise of internet-related tech companies, which eventually led to one of the largest stock market bubbles, the Dot Com Crash.

According to a study conducted a few years after the Dot Com Bubble, the NASDAQ Composite index – that includes numerous tech stocks – experienced excessive volatility between 1998 and 2001.

While the NASDAQ Composite includes all the equities that are listed on the exchange, the NASDAQ 100 – where tech stocks represented 70% of the total market cap in late 1998 – had a great influence on the index's valuation during the Dot Com Bubble. The chart above shows that the NASDAQ Composite's volatility had jumped as high as 53% in 1998, reaching 85% in 2001 when the tech stock market crashed.

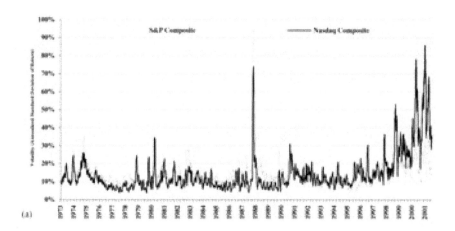

(a)

The NASDAQ Composite was so volatile at the time that its
volatility to the S&P 500 reached nearly 400% when the tech
bubble burst.

In the aftermath of the Dot Com Bubble, many tech companies had disappeared. Others had managed to survive the market crash. Since then, the technology sector has established a good

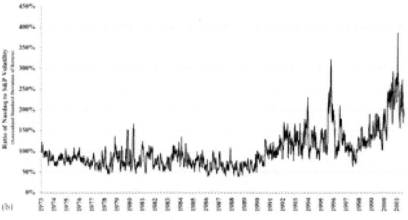

reputation and built out a decent infrastructure for itself.

As a result, the volatility of tech stocks has decreased significantly.

Between September 2009 and February 2020, the NASDAQ 100's volatility index (VXN) has mostly lingered between 15 and 20%, and never exceeding 50% (except in March 2020 due to the impact of the COVID-19 outbreak).

Speculation

The extent to which crypto traders base their actions on speculation also plays a part in driving Bitcoin's volatility.

At a surface level, when traders closely watch the markets, jumping between assets to buy the lows and sell the highs, potentially with leverage, without really investigating the fundamentals or looking at the broader context, this can make for irrational market behavior.

At a deeper level, if people hold, say, Bitcoin, only for speculative reasons – with no intention to use it as digital cash, or in the case of ETH as a means to develop decentralized applications, this can take away from liquidity (in extreme cases), but more importantly, such investors would be more likely to sell off their assets as soon as there is some bad news around it, if only because it is of no practical use to them.

So where do we stand with actual usage?
According to a Chainalysis report, only 1.3% of BTC transactions originated from merchants and 3.9% from (other) peer-to-peer (P2P) activities in the first four months of 2019, suggesting that speculation still remains the leading use-case for Bitcoin.

The Future of Bitcoin's Volatility

We know that Bitcoin is a highly volatile asset. But will it ever become less volatile?

If more people continue to adopt cryptocurrencies and the digital asset ecosystem continues to develop organically, then such would be expected.

If that happens, the crypto market's size could grow and present new real-world use-cases for consumers and businesses alike, while solving the speculation and liquidity issues of the digital asset economy.

Pro-crypto regulation could also help in lowering the magnitude of price swings as it would provide clarity, instill confidence and allow for much more capital from the regulated community to flow into the space.

Understanding the Various Ways to Invest in Bitcoin

Bitcoin was designed with the intent of becoming an international currency to replace government-issued (fiat) currencies. Since Bitcoin's inception in 2009, it has turned into a highly volatile investing asset that can be used for transactions where merchants accept it.

Could you and should you invest in Bitcoin? You can, and it depends on your appetite for risk. Learn the various types of ways you can invest in Bitcoin, strategies you can use and the dangers involved in this cryptocurrency.

Investment Types

Over the past decade, multiple ways to invest in Bitcoin have popped up, including Bitcoin trusts and ETFs comprised of Bitcoin-related companies.

Buying Standalone Bitcoin

The first way you can invest in Bitcoin is by purchasing a coin or a fraction of a coin via trading apps such as Coinbase. In most cases, you'll need to provide personal information to set up an account, then deposit money you'll use to purchase bitcoins. Some platforms may require a minimum deposit amount to purchase bitcoins. Then, as with any stock or ETF, you have access to Bitcoin's price performance and the option to buy or sell. When you buy, your purchase is kept safe in an encrypted wallet only you have access to.

Greyscale's Bitcoin Investment Trust (GBTC)

Investors looking to invest in Bitcoin through the capital markets can access an investment through Greyscale's Bitcoin Investment Trust (GBTC). Using Greyscale provides certain advantages that make an investment in bitcoin a more digestible option. For one, shares of GBTC are eligible to be held in certain IRA, Roth IRA, and other brokerage and investor accounts—allowing easy access for all levels of investors in a wide variety of accounts.

Investors are provided with a product that tracks the value of one-tenth of a bitcoin. As an example, if the value of Bitcoin is $1,000, each share of GBTC should have a net asset value of $100. This value is not without costs, as GBTC maintains a 2% fee that affects the underlying value.

In reality, investors are paying for security, ease of use, and liquidity (conversion to cash). By arranging strong offline storage mechanisms, GBTC allows investors who are less technical to access the bitcoin market safely.

GBTC trades on the capital markets as well, which allows it to trade at a premium or discount of its net asset value (NAV).

Amplify Transformational Data Sharing ETF (blok)

BLOK is an actively managed fund that has holdings in 15 different industries and is traded on the New York Stock Exchange Arca. The company invests in other companies that are involved with and developing blockchain technologies. BLOK's net expense ratio is 0.70%.

Bitwise 10 Private Index Fund

The Bitwise 10 Private Index Fund is based on the Bitwise 10 Large Cap Crypto Index, a basket of large capacity coins in which the company tries to provide security and the ease of use of a traditional ETF.

The Bitwise 10 Private requires a $25,000 minimum investment and has a fee ratio of 2.5%. Similar to GBTC, the assets are held in cold storage (offline), providing necessary security for its investors.

Investment Strategies

Hodl (an intentional misspelling of hold) is the term used in the bitcoin investment community for holding bitcoin—it has also turned into a backronym (where an acronym is made from an existing word)—it means "hold on for dear life." An investor that is holding their Bitcoin is "hodling," or is a "hodler."

Many people invest in Bitcoin simply by purchasing and holding the cryptocurrency. These are the people who believe in Bitcoin's long-term prosperity, and they see any volatility in the short term as little more than a blip on a long journey toward high value.

Long Positions on Bitcoin

Some investors want a more immediate return by purchasing Bitcoin and selling it at the end of a price rally. There are several ways to do this, including relying on the cryptocurrency's volatility for a high rate of return, should the market move in your favor. Several bitcoin trading sites also now exist that provide leveraged trading, in which the trading site effectively lends you money to hopefully increase your return.

Short Positions on Bitcoin

Some investors might bet on Bitcoin's value decreasing, especially during a Bitcoin bubble (a rapid rise in prices followed by a rapid decrease in prices). Investors sell their bitcoins at a certain price, then try to buy them back again at a lower price.

For example, if you bought a bitcoin worth $100, you would sell it for $100, and then wait for that bitcoin to decrease in value. Assuming the buyer of that bitcoin wanted to sell, you could buy it back at the lower price. You make a profit on the difference between your selling price and your lower purchase price.

It can be difficult to find a platform for short selling, but the Chicago Mercantile Exchange is currently offering options for Bitcoin futures.8

There is always the danger that the market will move against you, causing you to lose the money that you put up. Any trader should understand the concepts of leverage and margin calls before considering a shorting strategy.

Understanding Risk if You Invest in Bitcoin

Those fluctuations can be dramatic. In April 2013, the world gasped when Bitcoin's value jumped from around $40 to $140 in one month. That increase, however, paled in comparison to the Bitcoin surge of 2017. In January, Bitcoin was hovering between $900 and $1,000. In the first week of September, it pushed past $4,700, only to drop down near $3,600 two weeks later. By mid-December, it raced to an all-time high of $19,891.99, then plummeted to around $6,330 less than two months later.

Exchanges May Have Glitches and Hacks

Exchanges can be tricky because many of them have proven to be highly unreliable—especially in the early days of Bitcoin. One of the first and largest Bitcoin exchanges, Japan-based Mt. Gox, collapsed after being hacked—losing 850,000 bitcoins and hundreds of millions of dollars. In April 2016, a glitch in an exchange led to Bitcoin's price to momentarily drop to $0.60 on Coinbase.

The Bottom Line

Bitcoin's drawbacks aren't prohibitive. However, it is extremely important that you know what you're doing, and that you don't invest more than you can afford to lose. It is considered a very high-risk investment, meaning that it should represent a relatively small part of your investment portfolio.

If you are interested in investing in Bitcoin, you have multiple options. Buying bitcoins through an exchange subjects you to volatility, but opting for a trust or an ETF investing in crypto-tech companies could minimize the risk you'd face buying coins.

Lightning Source UK Ltd.
Milton Keynes UK
UKHW022259120421
381888UK00003B/445

9 781667 175058